LEMONS CAN POWER LIGHTBULBS.

TERMITES EVOLVED FROM COCKROACHES.

GIRLS SEND AND RECEIVE MORE THAN TWICE AS MANY TEXT MESSAGES AS BOYS.

Competitors race tricked-out portable potties at the annual outhouse races in Nevada, U.S.A.

Some Ice Age people used human skulls as drinking cups.

Russian supermarkets carry caviar-flavored potato chips.

The moon can look red during a lunar eclipse.

LAKE CHARGOGGAGOGGMANCHAUGG IN MASSACHUSETTS, U.S.A., HAS THE

Some **cobras** can spit venom up to **6.5 feet** (2 m)— that's longer than one hockey stick!

AGOGGCHAUBUNAGUNGAMAUGG, WORLD'S LONGEST NAME FOR A LAKE.

THE WORLD'S FASTEST TYPIST CAN REACH SPEEDS OF 200 WORDS PER MINUTE.

Porcupine quills were once used as toothpicks.

A puppy in California, U.S.A., was so tiny it was smaller than an iPhone!

AT TWO WEEKS OLD

THE ENERGY IN **ONE BOLT OF LIGHTNING** COULD TOAST ABOUT **100,000 SLICES OF BREAD.**

LILY PADS CAN GROW TO BE WIDER THAN A KING-SIZE BED.

You can buy lollipops made with flecks of real 24-karat gold.

THE WORLD'S HEAVIEST INSECT, THE giant weta, IS BIG ENOUGH TO EAT A CARROT.

WHEN A KID LOSES A TOOTH IN GREECE, HE THROWS IT ON THE ROOF!

THE WORLD'S RICHEST MAN IS WORTH MORE THAN $60 billion— that's enough to buy a fleet of a million sports cars.

Every 50 years, rocks from Mars fall to Earth.

LEECHES CAN BE USED TO PREVENT BLOOD CLOTS.

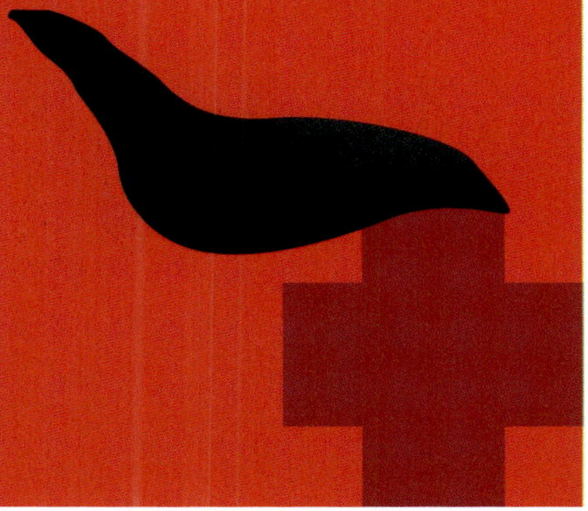

IT TAKES A **WET DOG** LESS THAN A **SECOND** TO SHAKE OFF **HALF** OF THE **WATER** ON ITS FUR.

PORTIONS OF ALBERT EINSTEIN'S BRAIN ARE ON DISPLAY IN A MUSEUM IN PHILADELPHIA, PENNSYLVANIA, U.S.A.

MUSICIANS PERFORMED A HIGH-FREQUENCY ROCK CONCERT IN AUSTRALIA THAT ONLY DOGS COULD HEAR.

A HORSE NEVER BREATHES THROUGH ITS MOUTH, EXCEPT IN EMERGENCIES.

Ants AS BIG AS THIS TOY CAR ONCE MARCHED ON EARTH.

Hummingbird nests are about the size of a golf ball.

Laid end-to-end, all of the **iPhones** and **iPads** **sold** in one year would stretch halfway around the **Earth.**

A SEA OTTER HAS A MILLION HAIRS ON ONE SQUARE INCH (6.5 sq. cm) OF ITS BODY.

EXTRACTS FROM HUMAN HAIR ARE USED TO MAKE SOME PIZZA CRUSTS.

There are eels in New Zealand that

The largest known snowflake was wider than a Frisbee.

THERE'S A WINGLESS ANT THAT CAN GLIDE FOR UP TO 90 FEET. (27 m)

A speck of blood contains about 5 million red blood cells.

can live for more than 100 years.

A FAMILY OF FIVE RODE A FIVE-SEAT BIKE 6,439 MILES (10,363 km) FROM KENTUCKY TO ALASKA, U.S.A., IN ONE YEAR.

COMPETITORS HURL THEIR OLD CELL PHONES FOR SPORT AT FINLAND'S ANNUAL MOBILE PHONE THROWING CHAMPIONSHIPS.

You can buy **socks** made from **bamboo.**

Some **gorillas** play **tag.**

TAG,
YOU'RE IT!

It's possible for a seed to blow across the ocean and sprout on another continent.

New Caledonian crows make right-handed and left-handed tools.

A NEW YORK ARTIST SELLS CLEAR CUBES filled with trash FOR AS MUCH AS $100.

TWO TEENAGERS **LAUNCHED** A LEGO MAN **85,000 FEET** (25,900 m) INTO **SPACE** USING A **HOMEMADE WEATHER BALLOON.**

ACTUAL VIDEO!

ASTRONAUT ALAN SHEPARD HIT **GOLF BALLS** WHILE ON THE MOON.

A mouse can squeeze through a **hole** the **size** of a piece of popcorn.

DOLPHINS EVOLVED FROM EARLY ANCESTORS OF GIRAFFES AND CAMELS.

PEOPLE IN THE UNITED STATES EAT MORE BANANAS THAN APPLES AND ORANGES COMBINED.

AN ARTIST CREATED A ONE-TON, LIFE-SIZE MAMMOTH OUT OF RECYCLED FARM EQUIPMENT.

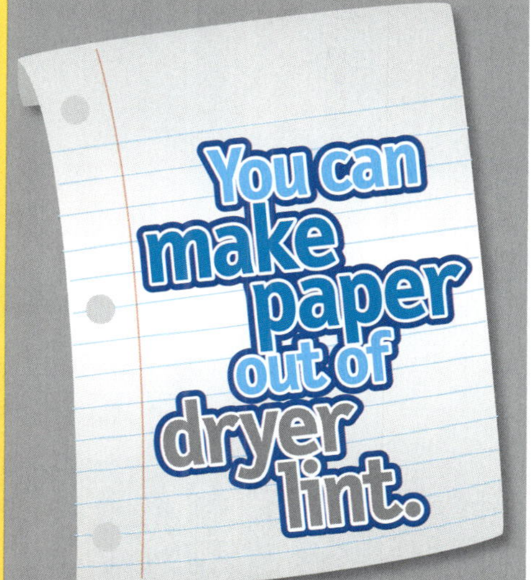

A giraffe sleeps 1½ hours a day.

14/KG/1